Volcanoes Earthquakes
Sudden Changes

Glen Phelan

Table of Contents

Develop Language............................2

CHAPTER 1   A Look Inside Earth.............4
                     Your Turn: Observe 9
CHAPTER 2   Volcanoes Change the Land......10
                     Your Turn: Predict.............15
CHAPTER 3   Earthquakes Change the Land....16
                     Your Turn: Interpret Data.......19

Career Explorations 20
Use Language to Restate 21
Science Around You..........................22
Key Words 23
Index 24

DEVELOP LANGUAGE

Almost 2,000 years ago, Mount Vesuvius **erupted**! It destroyed the town of Pompeii, Italy.

Look at the painting below. It shows an artist's idea of Mount Vesuvius in 79 A.D.

What do you notice in the painting?

I notice _____ .

Mount Vesuvius and Pompeii, Italy as they look today

Look at the photograph of Mount Vesuvius today. Discuss how the photo is different from the painting.

The painting shows _____ , but the photo shows _____ .

What do the pictures on this page show about volcanoes?

erupt – pour out ash and melted rock from inside Earth

lava

ash

painting of eruption in 79 A.D.

2 *Volcanoes and Earthquakes: Sudden Changes*

bodies of people preserved in mud and ash after the 79 A.D. explosion

CHAPTER 1
A Look Inside Earth

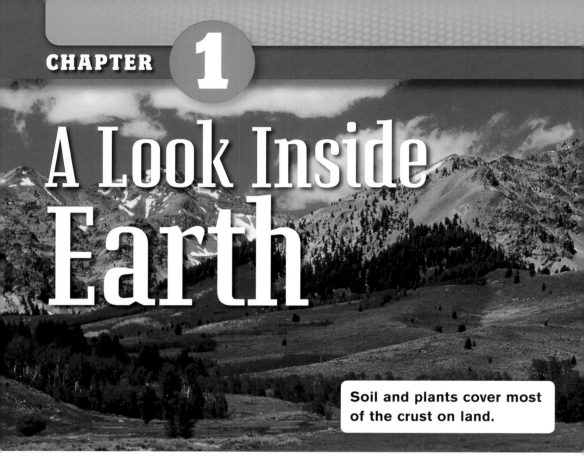

Soil and plants cover most of the crust on land.

Earth is made of layers. The top layer is the **crust**. The crust includes all of Earth's land and the ocean floors.

The crust is an average of 30 kilometers (about 18 miles) thick under land. It is even thicker under mountains. But this is thin compared to the other layers.

crust – the top layer of Earth

◀ Earth is made of three main layers.

Under the crust is a thick layer called the **mantle**. Most of the mantle is solid rock. But near the top is a layer of partly melted rock. This layer bends and flows very slowly.

Under the mantle is the **core**, the center part of Earth. The outer core is liquid metal. The inner core is a ball of solid metal.

mantle – the middle layer of Earth
core – the center part of Earth

KEY IDEA Earth is made up of layers.

Chapter 1: A Look Inside Earth

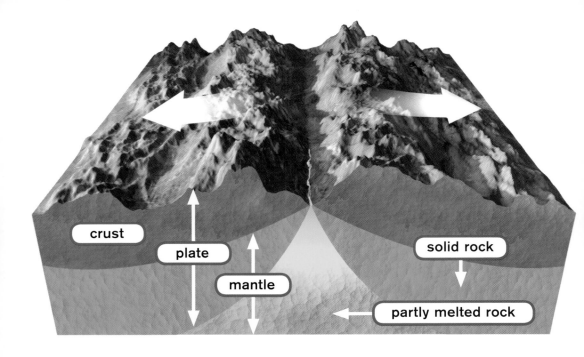

▲ Earth's plates float on the partly melted rock in the mantle.

A Broken Crust

The crust and the very top of the mantle form a solid, rocky shell around Earth. But this shell is not all one piece. It is broken into many sections called **plates.**

plates – large sections of Earth's crust and solid upper part of the mantle

Explore Language

Multiple-Meaning Words

A **plate** is a large section of Earth's crust and the top of the mantle.

A **plate** is also a flat dish that holds food.

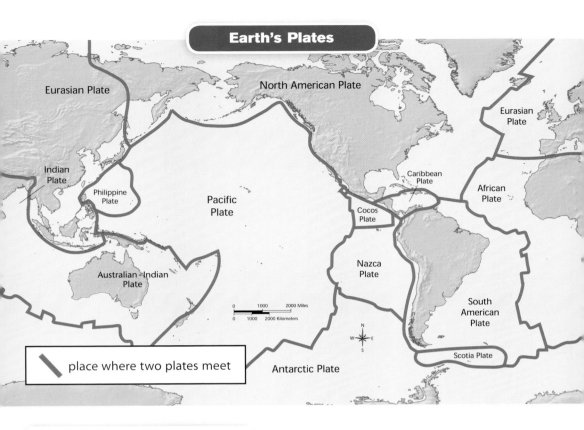

▲ Earth's surface is broken into about 20 plates.

Many of Earth's plates are huge. As the map shows, one plate includes almost the entire Pacific Ocean floor. Another plate includes most of North America and half of the Atlantic Ocean floor.

Chapter 1: A Look Inside Earth

Plates Move

Earth's plates move slowly over the partly melted rock in the mantle. The diagrams show three ways plates move.

Plates move only very short distances each year. But this slow movement can cause some big changes on Earth's surface.

KEY IDEA The crust is broken into sections called plates, which move slowly and change Earth's surface.

◀ Some plates spread away from each other.

▶ Some plates collide, or push against each other.

◀ Some plates slide past each other.

8 *Volcanoes and Earthquakes: Sudden Changes*

YOUR TURN

OBSERVE

Look at the picture below. The sand on both sides of the model fault is moving away from the fault. Now answer these questions.

1. Which picture on page 8 is like the picture of the sand fault?

2. What changes could this cause in Earth's surface?

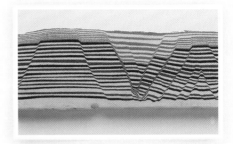

MAKE CONNECTIONS

You can compare Earth's layers to parts of a peach. What part of the peach would be the crust? What would be the mantle and the core?

USE THE LANGUAGE OF SCIENCE

What are three ways plates move?

Plates spread apart, collide, and slide past each other.

Chapter 1: A Layered Earth

CHAPTER 2

Volcanoes Change the Land

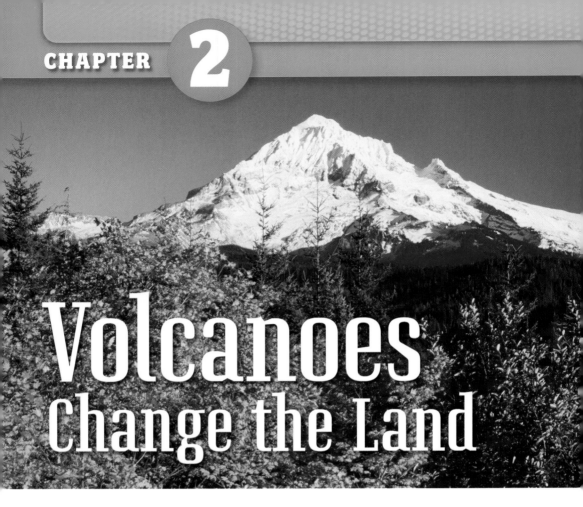

Sometimes, two plates that carry continents, or large areas of land, collide. The crust bends and crumples where the plates meet. Mountains form over millions of years as the plates slowly push against each other.

But what happens if one of the plates carries an ocean floor instead of a continent? Where they meet, the plate carrying an ocean floor sinks below the plate carrying a continent.

▲ Mount Hood is a volcano that formed when a plate carrying an ocean collided with a plate carrying a continent.

When a plate sinks below another plate, it moves into the hot mantle. That part of the plate becomes **magma**, or melted rock.

Sometimes the melted rock rises through cracks in the crust. If it pours onto the surface, the melted rock is called **lava**. The lava cools and becomes solid rock again. After many eruptions, lava can build up and make a mountain. A mountain formed in this way is called a **volcano**.

magma – melted rock under Earth's surface
lava – melted rock that reaches Earth's surface
volcano – a mountain made from melted rock that comes to Earth's surface and hardens

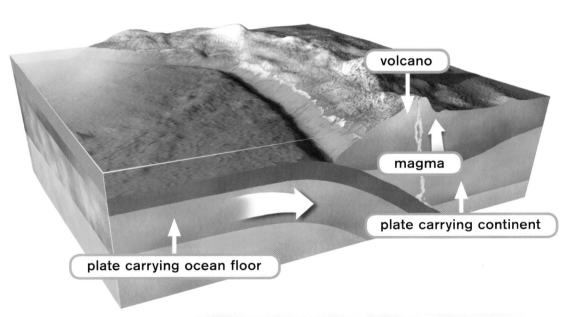

KEY IDEA Volcanoes form when melted rock reaches Earth's surface.

Chapter 2: Volcanoes Change the Land 11

Where Do Volcanoes Form?

Have you ever seen a volcano? The map shows where you can find them. Most volcanoes form at plate boundaries. A plate boundary is where two plates meet.

Look at all the volcanoes at the plate boundaries in the Pacific Ocean. This area has so many volcanoes that scientists call it the **Ring of Fire**.

Ring of Fire – area in the Pacific Ocean where many volcanoes form

▲ **Most volcanoes form at plate boundaries.**

12 *Volcanoes and Earthquakes: Sudden Changes*

Many volcanoes form on land. But most volcanoes form where you cannot see them—on the ocean floor.

Most volcanoes form where plates spread apart along the ocean floor. Magma rises between the plates and erupts as lava. The lava pushes the plates apart. The lava hardens into rock and builds an underwater chain of mountains called a mid-ocean ridge.

By The Way...

In 1963, a new island formed in the Atlantic Ocean. The island, Surtsey, formed when an underwater volcano built up enough to rise above the ocean's surface.

Volcanoes erupt in many places along a mid-ocean ridge.

Hot Spot Volcanoes

Some volcanoes form over an area called a **hot spot**, far away from plate boundaries. A hot spot is a place where magma rises from deep within the mantle. As a plate moves over a hot spot, the magma melts through the plate and erupts to form a volcano.

As the plate continues to move, the old volcano dies out. A new volcano forms over the hot spot. That's how the Hawaiian Islands formed.

hot spot – a place where magma rises from deep within the mantle

▼ **Lava pours out of Kilauea, a volcano in the Hawaiian Islands.**

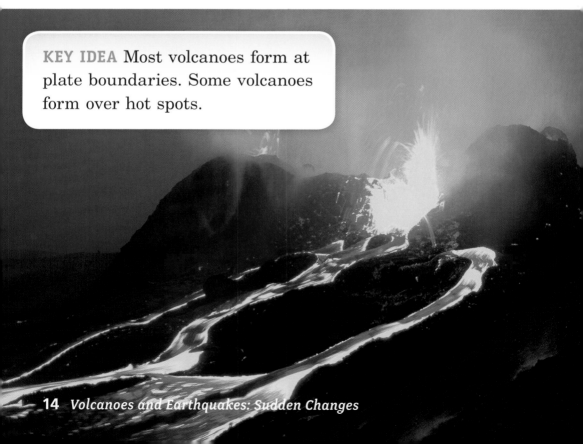

KEY IDEA Most volcanoes form at plate boundaries. Some volcanoes form over hot spots.

YOUR TURN

PREDICT

The drawing shows some of the Hawaiian islands. Find the hot spot under the Big Island of Hawaii. Notice how the plate has moved other islands away from the hot spot. Then answer the questions.

1. Why do you think the volcanoes on Maui no longer erupt?

2. Why do you think some of the volcanoes on Hawaii still erupt?

3. Make a prediction. Tell what could happen to the volcanoes of Hawaii as the plate keeps moving.

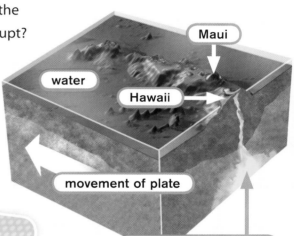

MAKE CONNECTIONS

Are there any volcanoes where you live? Look at the map on page 12 to find out. Where is the volcano closest to you?

 STRATEGY FOCUS

Determine Importance

Go over the information about volcanoes. What important ideas are presented? What details support those ideas?

Chapter 2: Volcanoes Change the Land

CHAPTER 3
Earthquakes Change the Land

Earth's plates do not move smoothly. The rough edges of the rock often get stuck against each other. The plates lock together.

Pressure builds as the plates try to move. Then suddenly the rock breaks free. The rock jerks when it suddenly moves. The sudden movement makes the ground shake. This shaking is an **earthquake**.

> **earthquake** – a shaking of the ground caused by the sudden movement of rock

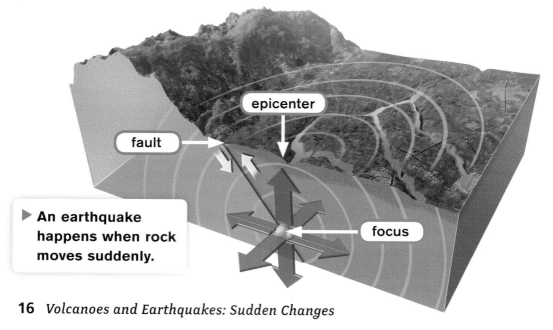

▶ An earthquake happens when rock moves suddenly.

16 *Volcanoes and Earthquakes: Sudden Changes*

Earthquakes happen along faults.

fault

Earthquakes happen along cracks called **faults**. The largest faults are where two plates meet. Smaller faults are between large blocks of rock in a plate.

faults – cracks in Earth's surface

Chapter 3: Earthquakes Change the Land

An earthquake raised the ground across this track and field.

How Earthquakes Change the Land

Hundreds of earthquakes happen every day. Most are so weak that people cannot feel them.

But stronger earthquakes can change the shape of the land. The shaking might raise part of the ground or make a crack. The shaking can damage or destroy buildings, bridges, and roads.

KEY IDEAS Earthquakes happen when rock moves suddenly along a fault. Earthquakes change the land.

YOUR TURN

INTERPRET DATA

The map below shows where earthquakes happened in the United States between 1975 and 1995. Look at the map with a friend and answer the questions.

1. Where did the most earthquakes happen?

2. Why do you think so many earthquakes happened in this part of the country? (Hint: Compare this map to the map on page 7.)

MAKE CONNECTIONS

Snap your fingers. Slowly press your thumb against your middle finger. Keep pressing until your fingers snap. How is this like an earthquake?

EXPAND VOCABULARY

The word **fault** is used in many contexts. Find out the meaning of **fault** in these sentences:

It is hard to find **fault** in your best friend.
I didn't do it! It's not my **fault**!

Write about the meaning of **fault** in each of these sentences.

Chapter 3: Earthquakes Change the Land

CAREER EXPLORATIONS

What Is a Geology Technician?

A geology technician helps geologists learn about rocks, volcanoes, and earthquakes.

- Read the chart. Would you like to be a geology technician?

- Explain your answer.

Would I like this career?	You might like this career if: • you like to explore outside. • you like to travel to different places.
What would I do?	• You would run machines that explore volcanoes and measure earthquakes. • You would help geologists do experiments and learn more about Earth's plates.
How can I prepare for this career?	• Learn about the kinds of rocks and land in your area. • To be a geology technician, you need to go to college. You would study math, geology, and chemistry.

USE LANGUAGE TO RESTATE

Words That Restate

When you restate, you say something in another way. You can start your restatement with the phrases **In other words** or **This means that**.

EXAMPLE

Sometimes the magma erupts. **This means that** it comes to Earth's surface.

With a friend, look at the captions in the book. Read the captions aloud. Then restate them in simpler language. Use **In other words** or **This means that**.

Write a Restatement

You learned how volcanoes change the land. Write a short report about this.

- Tell your reader how volcanoes form.
- Explain how they can change the land.
- Restate important ideas to help the reader understand.

Words You Can Use
in other words
this means that
another way to say this
or

SCIENCE AROUND YOU

ARE YOU PREPARED FOR AN EARTHQUAKE?

- Make sure bookcases are connected to walls.
- Make sure there is a strong table or desk that you can get under.
- Know how to shut off the gas if you smell a leak.
- Collect emergency supplies in a plastic trash can. Include these things:

  first-aid kit
  drinking water
  canned and dried food
  flashlights and batteries
  radio
  candles and matches

Posters remind people to be prepared if they live in places where earthquakes happen.

Look at the poster.

- Why is it important that bookcases be connected to walls?

- After an earthquake, fires can happen when gas escapes from broken gas lines. Which sentence in the poster helps to stop this damage?

- What else would you add to the poster?

Volcanoes and Earthquakes: Sudden Changes

Key Words

core the center part of Earth
Earth's **core** has solid and liquid metal.

crust the top layer of Earth
Earth's **crust** is very thin compared to Earth's other layers.

earthquake (earthquakes) a shaking of the ground caused by the sudden movement of rock
A strong **earthquake** can change the shape of the land.

fault a crack in Earth's crust
Earthquakes happen along **faults**.

hot spot (hot spots) a place where magma rises from deep within the mantle
The Hawaiian Islands formed above a **hot spot**.

lava melted rock that reaches Earth's surface
The volcano erupted **lava**.

magma melted rock under Earth's surface
Magma moves inside a volcano.

mantle the middle layer of Earth
Earth's **mantle** has a layer of partly melted rock near the top.

plate (plates) a large section of Earth's crust and solid upper part of the mantle
Most of the United States is on the North American **Plate**.

Ring of Fire an area in the Pacific Ocean where many volcanoes form
The **Ring of Fire** includes volcanoes that form along plate boundaries.

volcano (volcanoes) a mountain made from melted rock that comes to Earth's surface and hardens
Most **volcanoes** form where plates meet.

Index

core 5, 9
crust 4–6, 8–9
Earth 4–9, 16
earthquake 16–19, 22
erupt 2, 13–14
eruption 11
fault 17
hot spot 14–15
lava 11, 13
magma 11, 13–14
mantle 5–6, 9, 14
mid-ocean ridge 13
plate 6–14, 16–17
plate boundary 12, 14
pressure 16
Ring of Fire 12
volcano 11–15

MILLMARK EDUCATION CORPORATION
Ericka Markman, President and CEO; Karen Peratt, VP, Editorial Director; Rachel L. Moir, Director, Operations and Production; Mary Ann Mortellaro, Science Editor; Amy Sarver, Series Editor; Betsy Carpenter, Editor; Guadalupe Lopez, Writer; Kris Hanneman and Pictures Unlimited, Photo Research

PROGRAM AUTHORS
Mary Hawley; Program Author, Instructional Design
Kate Boehm Jerome; Program Author, Science

BOOK DESIGN Steve Curtis Design

CONTENT REVIEWER
Tom Nolan, Operations Engineer, NASA Jet Propulsion Laboratory, Pasadena, CA

PROGRAM ADVISORS
Scott K. Baker, PhD, Pacific Institutes for Research, Eugene, OR
Carla C. Johnson, EdD, University of Toledo, Toledo, OH
Donna Ogle, EdD, National-Louis University, Chicago, IL
Betty Ansin Smallwood, PhD, Center for Applied Linguistics, Washington, DC
Gail Thompson, PhD, Claremont Graduate University, Claremont, CA
Emma Violand-Sánchez, EdD, Arlington Public Schools, Arlington, VA (retired)

PHOTO CREDITS Cover © InterNetwork Media/Getty Images; 1 © Kevin Schafer/Peter Arnold, Inc.; 2-3 © O. Louis Mazzatenta/National Geographic/Getty Images; 2 © David A. Hardy/Photo Researchers, Inc.; 3a © Wojtek Buss/age fotostock; 3b, 7, 12 Maps by Mapping Specialists; 3c © Delderfield/age fotostock; 4-5 © William H. Mullins/Photo Researchers, Inc.; 5 © Roger Harris/Photo Researchers, Inc.; 6, 8a, 8b, 8c, 11, 15, 16 Illustrations by Chuck Carter; 9a © Clive Streeter/Dorling Kindersley; 9b © Clearvista Photography/Alamy; 9c and 9d Lloyd Wolf for Millmark Education; 10 © Design Pics Inc./Alamy; 13a and 23 © Arctic Images/Alamy; 13b © Cousteau Society/Getty Images; 14 and 21 © Photo Resource Hawaii/Danita Delimont; 17 © James Balog/Getty Images; 18 © AP Images/Wang Yuan-mao; 19 © U. S. National Earthquakes Information Center; 20 © AP Images/Ted S. Warren; 22a © Shapiso/Shutterstock; 22b © Mike Grindley/Shutterstock; 24 © Digital Vision/Punchstock

Copyright © 2008 Millmark Education Corporation

All rights reserved. Reproduction of the whole or any part of the contents without written permission from the publisher is prohibited. Millmark Education and ConceptLinks are registered trademarks of Millmark Education Corporation.

Published by Millmark Education Corporation
7272 Wisconsin Avenue, Suite 300
Bethesda, MD 20814

ISBN-13: 978-1-4334-0085-8
ISBN-10: 1-4334-0085-5

Printed in the USA

10 9 8 7 6 5 4 3 2 1